ASTRONOMY JAMES E. WEBB SPACE TELESCOPE

More Details About James Webb Space Telescope

Table Of Content

Chapter 1
Chapter 2
Chapter 3

Chapter 1

All About James E. Webb Space Telescope

From 1949 to 1952, James Edwin Webb—a member of the American government who died on March 27, 1992—held the position of Undersecretary of State. In the period from February 14, 1961, until October 7, 1968, he served as NASA's second administrator.

Webb oversaw each of the crucial first human flights throughout the Mercury and Gemini projects up to a few days before the launch of the first Apollo mission. He served as NASA's administrator from the start of the Kennedy administration until the conclusion of the Johnson administration. He handled the Apollo 1 fire very well.

William E. Webb

Younger years and personal lives

In the North Carolina county of Granville, in the hamlet of Tally Ho, Webb was born in 1906. His father oversaw the public schools in Granville County. [2] At the University of North Carolina in Chapel Hill, he finished his undergraduate studies, earning a Bachelor of Arts in Education there in 1928. He belonged to the fraternity Acacia.

From 1930 to 1932, Webb was an active duty Marine Corps pilot after being promoted to the rank of second lieutenant. After that, Webb pursued legal studies at The George Washington University Law School, where he graduated with a J.D. in 1936. He was admitted to the District of Columbia Bar the same year.

In 1938, Webb wed Patsy Aiken Douglas; the couple had two kids. As a Freemason, he was. [3]

Career

Staff of the US House of Representatives

In Washington, D.C., Webb started his lengthy career in public service by working as US Representative Edward W. Pou of North Carolina's secretary from 1932 to 1934. Pou served as both the House's dean and the head of the Rules Committee.

Pou had a significant impact on the passage of the first New Deal legislation during the first 100 days of Franklin D. Roosevelt's presidency with the help of Webb. In addition to his clerical tasks, Webb helped the ill and elderly Pou physically.

Support staff for a private attorney

From 1934 to 1936, Webb worked as an assistant in the law office of Oliver Max Gardner, a friend of President Roosevelt who was also a former governor of North Carolina. Gardner helped Webb complete his legal education.

The government stopped commercial airplane businesses from transporting airmail during the 1934 Air Mail Scandal. Thomas Morgan, the president of the Brooklyn-based Sperry Gyroscope Company, headed a group of airline executives who engaged Gardner's office to defend them. Contracts with private airlines resumed as a consequence of the successful settlement.

Sperry Gyroscope's director of human resources

Following their discussions, Sperry Gyroscope recruited Webb to serve as the company's personnel director and Thomas Morgan's assistant. Webb served as Sperry's secretary-treasurer from 1936 to 1944 before rising to the position of vice president. During his leadership, Sperry increased its workforce from 800 to more than 33,000, and during World War II, it emerged as a significant provider of airborne radar systems and navigational aids. [6]

Re-enlistment of Marines

Webb initially intended to rejoin the Marines, but because of the significance of his job at Sperry to the war effort, he was postponed. On February 1st, 1944[7], he rejoined the Marine Corps, and within a short period of time, first as a captain and subsequently as a major, he was in charge of Marine Air Warning Group One, the 9th Marine Aircraft Wing. Henry Gorham Webb, another Marine Corps commander who had served with VMF-211 at the Battle of Wake Island and had since been captured, was at the time a prisoner of war in Japan.

He was given responsibility for a radar operation for the invasion of mainland Japan. His instructions were delayed, and the surrender of Japan on September 2, 1945, prevented him from seeing any action. He was supposed to travel for Japan on August 14, 1945, but his orders were delayed.

Office of Budget

After the war, Webb relocated to Washington, DC, where he briefly worked as Gardner's executive assistant before being appointed head of the Bureau of the Budget in the Office of the President of the United States, a post he maintained until 1949. Gardner and Treasury Secretary John Snyder made Webb Harry S. Truman's recommendation for the position. Webb's nomination was seen as subordinating the BoB to the Treasury because of his connection to the Treasury Department. He hadn't been informed of the final decision to nominate him, which is why Webb was shocked by his selection.

Each year, the President's proposed budget was produced by the Bureau of the Budget for presentation to Congress. After the significant outlays of World War II, Truman's goal for the budget was to balance it.

Department of StateEdit

The U.S. Department of State then appointed Webb as an undersecretary of state, a position he started in January 1949. Reorganizing the Department with the addition of 12 new Presidential appointees and a reduction in the authority of subordinate officials was Webb's first task from Secretary Dean Acheson. Additionally, Webb organized the secretariat's flow of intelligence and information related to foreign policy. The Department, which had been losing authority and influence to the military, increased its links to the President when the new structure became law in June 1949.

The Department of State at the time debated whether or not military action would be required to limit the Soviet Union if merely diplomatic measures failed. In his capacity as Director of Policy Planning, Paul Nitze argued for a military escalation of NATO troops in a confidential paper titled NSC 68. Although Louis A. Johnson, the secretary of defense, opposed an increase in the defense budget, Webb was able to persuade Truman to adopt the NSC 68 recommendations.

The North Korean Army invaded South Korea on June 25, 1950. Three suggestions were made by Webb and Secretary Acheson:

involve the UN, dispatch the Pacific Fleet of the Navy into the Yellow Sea, and order an airstrike on the Korean tanks. The first two suggestions were put into action right away by Truman, but he put off using force for a few days.

The lack of US readiness was attributed to the Defense Department, and Johnson made an effort to place the responsibility on Acheson. George Marshall was brought out of retirement to take over as the new Secretary of Defense after Webb and others used his connections in Congress to persuade Truman to take over for Johnson.

In order to strengthen American skills in psychological warfare, Webb formed Project Troy in 1950 with the assistance of academic researchers. This project focused on finding ways to get over Soviet efforts to jam Voice of America transmissions.

Webb's influence diminished as the Department's emphasis shifted to the Korean War.

[Reference required] Paul Nitze, director of policy planning at the State Department, was the architect of NSC 68 and later served as Secretary Acheson's primary counselor. After a disagreement, Nitze publicly called for Webb's resignation, but the issue was finally resolved. [Reference required] In 1952, after developing migraines, Webb resigned.

Webb moved from Washington to Oklahoma City to work for the Kerr-McGee Oil Corp., but he continued to be involved in government affairs, sitting on the Draper Committee in 1958, for example.

NASA

On February 14, 1961, Webb accepted President John F. Kennedy's nomination to lead NASA as administrator, succeeding Deputy Administrator Hugh L. Dryden, who had served as acting administrator. The Apollo program was overseen by Webb while NASA worked to achieve Kennedy's objective of putting an American on the moon by the end of the 1960s. After Kennedy announced the objective of a human lunar landing on May 25, 1961, Webb pushed for NASA in Congress for seven years until he departed NASA in October 1968. As a veteran

He was able to secure ongoing funding and support for Apollo because of his insider status in Washington and the support of President Lyndon B. Johnson.

NASA transformed under Webb's leadership from a dispersed network of research institutes to a cohesive institution. He played a significant part in founding Houston's Manned Spacecraft Center, which ultimately became the Johnson Space Center. Webb made sure that NASA carried out a program of planetary exploration with the Mariner and Pioneer space missions against demands to concentrate on the Apollo program.

Webb informed the reporters after the 1967 Apollo 1 mishap, "We've always anticipated that sooner or later, something similar will occur. Who would have imagined that a catastrophe would occur first?" According to a protocol put in place after the in-flight disaster on Gemini 8, Webb went to

Johnson and requested that NASA be permitted to manage the accident investigation and oversee its recovery (1966).

He pledged to be forthright when assigning responsibility for Apollo 1—including, if necessary, to himself and NASA management. The organization set out to learn more about the disaster, address issues, and go on with plans for the Apollo 11 lunar landing.

When presenting the inquiry board's findings to different congressional committees, Webb almost always accepted responsibility for his actions. Whether by chance or on purpose, Webb was able to divert some of the criticism of the tragedy away from the Johnson administration and NASA as a whole. Consequently, NASA's reputation and public support were substantially unharmed.

Johnson and Webb had a good relationship as Democrats, but when Johnson decided not to seek reelection, Webb decided to resign as administrator so that Richard Nixon, the Republican who would become the next president, could choose a replacement.

Webb instructed NASA to have Apollo 8 ready for a potential lunar orbital mission in 1968 after learning through CIA sources that the Soviet Union was preparing its hefty N1 rocket for a human lunar trip. The N-1 was called "Webb's Giant" at the time because several individuals at the time questioned Webb's claims about the Soviet Union's capabilities.

 But since the fall of the Soviet Union, information concerning the Soviet Moonshot has supported Webb's assertion. Just before the first human mission in the Apollo program, Webb departed NASA in October 1968.

In Space Age Management: The Large-Scale Approach (1969), Webb portrayed the space program as a successful administration model that could be expanded to meet significant social issues. Webb drew on his experience working for NASA in this publication.

Johnson gave Webb the Presidential Medal of Freedom in 1969. The Smithsonian Institution awarded Webb the Langley Gold Medal in 1976 as well.

Legacy

In the 1998 miniseries From the Earth to the Moon, Dan Lauria portrayed Webb.

Ken Strunk portrayed Webb in the 2016 movie Hidden Figures.

The Next Generation Space Telescope, which was the original name of NASA's James Webb Space Telescope (JWST), was renamed in Webb's honor in 2002. It will be launched on December 25, 2021, and is referred to be the Hubble Space Telescope's replacement.

Dispute about the naming of the telescope

Following State Department regulations set in place in 1947, Webb served in a senior position during the so-called lavender panic, which resulted in the termination of hundreds of gay employees from the organization, from 1950 to 1952. According to the records, President Truman and Webb met on June 22, 1950, to discuss how the White House, State Department, and Hoey Committee might "work together on the homosexual investigation." Truman agreed to send two White House aides with Webb to meet with the Hoey Committee to establish a working method.

Throughout Webb's time at the State Department, gay state employees continued to be fired; from 1950 to 1952, hundreds of homosexual employees were said to have been fired by Webb's colleagues.] I don't regard [Webb] as having any type of leadership role in the lavender panic, according to historian David K. Johnson, who claims that Webb's presence at the White House meeting was in the context of limiting the hysteria that members of Congress were stoking up.

A Scientific American essay published in March 2021 requested NASA to rename the James Webb Space Telescope since Webb was allegedly involved in the State Department's effort to remove homosexuals from the government workforce.

The press covered this topic extensively. Scientists who objected to the telescope being named in honor of Webb cited the instance of NASA budget analyst Clifford Norton, who was detained and sacked after being accused of homosexual activity in 1963. NASA described Norton's alleged behavior as "immoral, disgusting, and shameful." The Deputy Administrator of NASA, Robert Seamans, was in charge of personnel problems; there is no concrete proof that Webb was aware of Norton's termination.

Perhaps such terminations were "custom inside the agency" at the time. There is no proof that Webb oversaw any persecution or had "any type of leadership role in the lavender fear," according to historian David K. Johnson, author of the 2004 book The Lavender Scare. The early charges that Webb was involved in the lavender panic, according to astronomer Hakeem Oluseyi, were based on a comment by John Peurifoy (who, like Webb, had the title of "Undersecretary of State") that was incorrectly ascribed to Webb.

An intern working with NASA's chief historian Brian Odom and NASA Communications Specialist Catherine Baldwin wrote in an email acquired by Nature under the Freedom of Information

Act (FOIA) in March 2022 that "[t]hat Webb took a prominent role in the Lavender Scare is obvious."

The intern's findings are disputed by other sources. The New York Times reported in March 1952, shortly after Webb departed the Department of State, that 126 federal employees had been let go. The assertion that the termination of LGBTQ employees stopped when Webb departed State is refuted by the statistics, which show that by April 1953, that figure had doubled as 425 employees were let go.

President Dwight D. Eisenhower issued Executive Order 10450 in April 1953, approximately a year after Webb departed the State Department. This order significantly expanded the Lavender Scare campaign and resulted in hundreds of dismissals. In 2021, David K. Johnson, the author of the book the intern quoted, had said to the Washington Post that "he knew of no proof that Webb had a key role in the initiative."

After conducting a study and determining that there was "no evidence at this moment that justifies altering the name," NASA said on September 30, 2021, that it will maintain the JWST moniker.

That administrator Webb to "be held liable for that action when there's no evidence to even indicate [that he participated in it] is an injustice," according to former administrator Sean O'Keefe, who decided to name the telescope after administrator Webb.

How did he function at NASA?

President Kennedy wanted to send an American to the moon by 1970, and Webb oversaw NASA's efforts to achieve that goal.

Despite his initial reluctance to accept the position, he eventually rose to the position of NASA administrator, guiding the agency to become a major pioneer in space research from a very small collection of aerospace organizations.

Regarding Kennedy's offer, Webb stated: "I believed that I had established a pattern for my life and that, in any case, I was not the most suitable candidate. Someone who understood more about space exploration and rocketry, in my opinion, would be a better person."

Soon after Webb was appointed, in May 1961, Kennedy declared a human trip to the moon by October 1968.

Due to his political contacts and oversight of the Apollo mission, Webb was able to secure ongoing Congressional funding for the NASA program.

Under Lyndon B. Johnson, he persisted in his position, and NASA honored him as a leader who was able to prioritize research and technology above human spaceflight.

Since the Soviet Union likewise intended to be the first to send people into space, the US space agency claims that it depoliticized the space race.

In early meetings between himself, President Kennedy, and Vice President Johnson, as recorded in Webb's transcripts, he reportedly said to them: "And so far as I'm concerned, I'm not going to run a program that's simply a one-shot program.

"It will be a balanced program that serves the needs of the nation if you want me to be the administrator,"

In line with Webb's plan, NASA continued to invest in the creation of robotic spacecraft, which subsequently helped humans explore the lunar surface and inspired the launch of Mars and Venus research missions.

With the help of his mission, the US was also expected to offer the first-ever glimpse of the surface of space, and by 1965, he had already planned the launch of a large telescope into orbit. Hubble would subsequently develop from this.

Despite his commitment to the moon mission, he left his position at NASA only a few months before the July 1969 lunar landing.

In a statement honoring Webb, NASA noted that the original probe launches he oversaw resulted in "more than 75 space research missions to investigate the stars and galaxies, our own Sun, and the as-yet-unknown environment of space above the Earth's atmosphere."

The most prosperous era of astronomical discovery in history, which is still going strong now, was laid out by missions like the Orbiting Solar Observatory and the Explorer series of astronomical satellites, it was stated.

After NASA, what did he do?

After leaving NASA, Webb served on several advisory boards, including one for the Smithsonian Institution as a regent.

He received the Sylvanus Thayer Award for his devotion to his nation in 1981 from the United States Military Academy at West Point.

In March 1992, a heart attack claimed his life.

Who gave the telescope its name, James Webb?

Since the turn of the century, the James Webb telescope has been in development; it was given its official name in 2002.

Despite efforts by certain US scientists to have the name changed because they thought he was against LGBTQ+ equality in the NASA workforce, no such proof was shown, hence it was decided to keep the name.

The telescope's name was revealed by NASA Administrator Sean O'Keefe, who said: "It is appropriate that Hubble's successor be called in honor of James Webb. He made it possible for us to see the spectacular scenery of space for the first time.

"He led our country on its first exploratory missions, bringing our dreams to life. He helped NASA lay the groundwork for one of the most productive eras of cosmic research.

Therefore, with the aid of the Hubble Space Telescope, the Chandra X-ray Observatory, and the James Webb Telescope, we are now revising the textbooks.

Additional actions after

What caused James Webb to leave just before Apollo 8?

He has been in charge of NASA's development for the human lunar missions since 1961. But for whatever reason, just days before the Apollo 8 mission to the Moon, which marked the conclusion of eight years of labor, he announced his resignation as NASA Administrator. What would make him do this?

Webb told the reporters after the Apollo 1 disaster in 1967, "We've always known that something like this was going to happen sooner or later. Who would have imagined that a disaster would occur first on the ground? President Johnson was contacted by Webb, who requested that NASA be given control over the disaster investigation and recovery efforts.

He vowed to allocate responsibility fairly and to the right parties, including himself and NASA management. The organization set out to learn more about the disaster, address issues, and go on with plans for the Apollo 11 lunar landing.

Webb, a Democrat closely associated with Johnson, decided to resign as administrator after Johnson announced he would not seek reelection. This would have allowed the next president to choose his administrator.

Telescope in space

A telescope in space used to examine celestial objects is known as a space telescope or space observatory. The American Orbiting Astronomical Observatory, OAO-2 launched in 1968, and the Soviet Orion 1 ultraviolet telescope on space station Salyut 1 in 1971 was the first operational telescope, both of which were proposed by Lyman Spitzer in 1946. Space telescopes do not experience the light pollution that ground-based observatories do, nor do they experience the filtering and distortion (scintillation) of the electromagnetic radiation they detect.

The two main categories of satellites are those that map the whole sky (astronomical surveys) and those that concentrate on certain astronomical objects or regions of the sky and beyond. Space telescopes are separate from Earth-imaging satellites, which are used for weather monitoring, espionage, and other sorts of information collection. These spacecraft look toward Earth for satellite imaging.

Advantages

The filtering and distortion of electromagnetic radiation (scintillation or twinkling) caused by the atmosphere place restrictions on the ability to conduct astronomy from ground-based observatories on Earth.

A telescope circling Earth outside of the atmosphere is not affected by Earth's artificial light pollution or twinkling. As a consequence, a space telescope's angular resolution is often significantly greater than a ground-based telescope's with a comparable diameter. However, many bigger terrestrial telescopes use adaptive optics to lessen atmospheric impacts.

For frequency ranges beyond the optical window and the radio window, the only two wavelength ranges of the electromagnetic spectrum that are not significantly affected by the atmosphere, space-based astronomy are increasingly crucial. For instance, it is practically hard to conduct X-ray astronomy from Earth, and orbiting X-ray telescopes like the Chandra X-ray Observatory and the XMM-Newton observatory are the sole reason it has attained its present significance in astronomy. UV and infrared radiation are also substantially refracted.

Disadvantages

Building costs for space telescopes are much higher than for terrestrial telescopes. Space telescopes are also very challenging to maintain because of their remoteness. The Space Shuttle was used to maintain the Hubble Space Telescope, but most space telescopes cannot be maintained at all.

Space observatories' potential

NASA, ISRO, ESA, CNSA, JAXA, and the Soviet space program, which was subsequently superseded by Russia's Roscosmos, have all launched and managed satellites. Many space observatories have already finished their missions as of 2018, however, others continue to operate for lengthy periods. However, timely and enough financing is necessary for space telescopes and observatories to be made available in the future. While NASA, JAXA, and the CNSA are planning new satellite observatories, there may be coverage gaps that are not immediately filled by these programs, which might have an impact on research in basic science.

The James Webb Space Telescope (JWST) is an observatory that was built mainly for infrared astronomy. Its enormously enhanced infrared resolution and sensitivity enable it to observe objects that are too early, far away, or dim for the Hubble Orbit

Telescope since it is the greatest optical telescope in space. This is anticipated to open up a wide variety of astronomical and cosmological inquiries, including the observation of the earliest stars and the development of the first galaxies, as well as the comprehensive atmospheric characterization of exoplanets that may harbor life.

Chapter 2

Exoplanet

A planet beyond the Solar System is referred to as an exoplanet or extrasolar planet. In 1917, the first potential sign of an exoplanet was observed, but it was not taken seriously. In 1992, the first detection confirmation took place. 1988 saw the discovery of a different planet, which was verified in 2003. There are 3,779 planetary systems with 3,108 confirmed exoplanets as of 1 July 2022, 826 of which have multiple planets.

Exoplanets may be found using a variety of techniques. The majority of exoplanets have been discovered using transit photometry and Doppler spectroscopy, but these techniques suffer from a glaring observational bias that favors the discovery of planets close to the star; as a result, 85% of exoplanets discovered are located within the tidal locking zone.] Multiple planets have been seen around a star in several instances.

There is an "Earth-sized"[b] planet in the habitable zone for around 1 in 5 Sun-like stars. It is possible to predict that there are 11 billion potentially hospitable Earth-sized planets in the Milky Way if there are 200 billion stars in the galaxy. This number rises to 40 billion if planets circling the abundant red dwarf stars are taken into account.

Draugr, which is about twice as large as the Moon and is also known as PSR B1257+12 A or PSR B1257+12 b, is the least massive planet that is currently understood. The planet HR 2562 b, which has a mass almost 30 times that of Jupiter, is the most massive one identified in the NASA Exoplanet Archive.

Some definitions of a planet, however (based on the nuclear fusion of deuterium), state that it is too big to be a planet and maybe a brown dwarf. For exoplanets that are nearest to their star, the known orbital periods range from a few hours to thousands of years. It may be difficult to determine whether certain exoplanets are gravitationally connected to a star since they are so distant from the star.

The Milky Way contains almost all of the planets that have been discovered so far. Extragalactic planets, or exoplanets that are located in galaxies outside of the local Milky Way galaxy, may exist, according to some evidence. The closest exoplanets are 4.2 light-years (1.3 parsecs) away from Earth and revolve around Proxima Centauri, the Sun's nearest neighbor.

The quest for alien life has gained more attention as a result of the finding of exoplanets. The habitable zone, often known as the "Goldilocks zone," is the region where planets may circle stars and support liquid water on their surfaces, which is necessary for the existence of life as we know it. However, a variety of other characteristics are also taken into account by the study of planetary habitability when deciding whether a planet is suitable for supporting life.

Rogue planets have no known star in their orbit. Particularly if they are gas giants, which are sometimes classed as sub-brown dwarfs, these objects are regarded as belonging to a distinct type of planet.] There may be billions or more rogue planets in the Milky Way.

Definition

Exoplanets are not included in the International Astronomical Union's (IAU) official definition of the word "planet," which exclusively refers to objects in the Solar System.

A working definition of "planet" was included in a position statement that the IAU Working Group on Extrasolar Planets released in 2001 and was updated in 2003.

The following criteria were used to determine what an exoplanet was:

"Planets" are celestial bodies that circle other stars or stellar remnants and have real masses below the thermonuclear fusion threshold for deuterium, which is presently estimated to be 13 Jupiter masses for objects with solar metallicity (no matter how they formed). Extrasolar objects should have a minimum mass/size that is equivalent to that of planets in our solar system.

No matter how or where they evolved, substellar objects with actual masses greater than the limiting mass for deuterium thermonuclear fusion are referred to as "brown dwarfs".

Sub-brown dwarfs, not planets, are free-floating objects in newborn star clusters with masses below the limiting mass for thermonuclear fusion of deuterium (or whatever name is most appropriate).

In August 2018, the IAU's Commission F2: Exoplanets and the Solar System modified its working definition.

The following is the current official working definition of an exoplanet:

"Planets" are objects that orbit stars, brown dwarfs, or stellar remnants and have true masses below the limiting mass for thermonuclear fusion of deuterium, which is currently estimated to be 13 Jupiter masses for objects of solar metallicity. These objects also have mass ratios with the central objects that are below the $L4/L5$ instability (M/Central 2/(25+621). (no matter how they formed).

It should be the same as the criteria used in our Solar System to define a planet's minimal mass and size for extrasolar objects.

Alternatives

Not always is the working definition of the IAU applied. A different recommendation is to identify planets from brown dwarfs based on how they formed. Giant planets are hypothesized to originate from core accretion, which may sometimes result in planets with masses over the deuterium fusion threshold; large planets of this kind may have previously been seen. Brown dwarfs are objects that are below the 13 MJup limits and may even be as low as 1 MJup. They arise similarly to stars through the direct gravitational collapse of gas clouds.

The atmospheres of objects in this mass range that circle their stars at great distances of hundreds or thousands of AU and have enormous star/object mass ratios are most likely brown dwarfs, as opposed to planets created via accretion, which would have higher abundances of heavier elements. As of April 2014, the majority of clearly observable planets are huge and have broad orbits, which likely reflect the low-mass end of the brown dwarf formation. According to one research, objects larger than 10 MJup did not originate as planets but rather as a result of gravitational instability.

The 13 Jupiter mass threshold is also not physically significant in a precise way. There are certain things with mass below that limit where deuterium fusion can take place. The composition of the item has some bearing on how much deuterium fuses. The Extrasolar Planets Encyclopaedia said that objects up to 25 Jupiter masses were included as of 2011 and that the lack of a distinctive signature in the reported mass spectrum around 13 MJup "reinforces the decision to overlook this mass restriction."

Based on the research on mass-density connections, this limit was raised as of 2016 to 60 Jupiter masses. The 13 Jupiter-mass distinction by the IAU Working Group is physically unjustified for planets with rocky cores and observationally problematic owing to the sin I ambiguity, according to the Exoplanet Data Explorer, which covers objects up to 24 Jupiter masses. Objects having a mass (or minimum mass) equal to or lower than 30 Jupiter masses are included in the NASA Exoplanet Archive.

Instead of the mechanism or location of deuterium fusion, another factor used to distinguish between planets and brown dwarfs is whether the core pressure is dominated by coulomb pressure or electron degeneracy pressure, with the dividing line falling at around 5 Jupiter masses.

Nomenclature

The International Astronomical Union approved the protocol for naming exoplanets as an extension of the convention for naming multiple-star systems (IAU). The IAU designation for an exoplanet circling a single star is created by adding a lowercase letter to the star's designated or proper name. The letters are assigned in the sequence in which the planets were found orbiting their parent stars, starting with "b" for the first planet identified in a system (the parent star is regarded as "a") and continuing with "b" for each additional planet.

The nearest planet to the star receives the next letter, followed by the remaining planets in order of orbital size if many planets in the same system are found at the same time. There is a temporary IAU-approved standard that allows for the designation of circumbinary planets. There are just a few exoplanets with IAU-approved proper names. There are more naming schemes.

Background of detection

Although extrasolar planets have been hypothesized to exist for millennia, there is no way to confirm their existence, estimate their frequency, or determine how similar they could be to the planets of the Solar System. Astronomers disproved several detection claims that were made in the eighteenth century.

The first indication of a potential exoplanet was seen circling Van Maanen 2 in 1917, although it was not acknowledged as such. Using the 60-inch telescope at Mount Wilson, the astronomer Walter Sydney Adams, who ultimately rose to the position of director of the observatory, created a spectrum of the star. He believed the spectrum to be that of an F-type main-sequence star, but it is now believed that such a spectrum might be the consequence of the debris left behind when a nearby exoplanet was crushed into dust by the star's gravity and the material subsequently fell onto the star.

In 1988, there was the first alleged exoplanet discovery by science. A few years later, in 1992, the finding of numerous planets of terrestrial masses circling the pulsar PSR B1257+12 provided the first confirmation of the first detection. [40] The discovery of a huge planet in a four-day orbit around the neighboring star 51 Pegasi in 1995 provided the first evidence of an exoplanet circling a main-sequence star. Telescopes have photographed a small number of exoplanets directly, but the great majority have been discovered via indirect techniques like the transit method and the radial-velocity approach.

Using the Chandra X-ray Observatory and a method for finding planets called microlensing, scientists discovered proof of planets in a far-off galaxy in February 2018 "These exoplanets range in size from as huge as Jupiter to as (relatively) tiny as the moon. Contrary to Earth, most exoplanets are loosely circling between stars or wandering through space since they are not strongly connected to their stars. We can calculate that there are more than a trillion planets in this [far-off] galaxy."

The 5000th exoplanet outside of our solar system was verified on March 21, 2022.

Earliest theories

Italian philosopher Giordano Bruno advanced the idea that fixed stars are analogous to the Sun and are also accompanied by planets in the sixteenth century. Bruno was an early proponent of the Copernican hypothesis that the Earth and other planets circle the Sun (heliocentrism).

A similar notion was raised by Isaac Newton in the "General Scholium" that wraps up his Principia in the seventeenth century. He added, "And if the fixed stars are the centers of comparable systems, they will all be created according to a similar pattern and subject to the sovereignty of One," drawing a connection to the planets of the Sun.

Otto Struve wrote that there is no compelling reason why planets could not be much closer to their parent stars than is the case in the Solar System in 1952, more than 40 years before the first hot Jupiter was discovered. Struve also suggested that Doppler spectroscopy and the transit method could detect super-Jupiters in short orbits.

Disputed assertions

Exoplanet detection claims date back to the eighteenth century. The binary star 70 Ophiuchi is a part of some of the earliest. At the Madras Observatory of the East India Company, William Stephen Jacob wrote in 1855 that orbital irregularities proved the existence of a "planetary body" in this system "very plausible."]

According to Thomas J. J. See of the University of Chicago and the United States Naval Observatory, the orbital anomalies in the 70 Ophiuchi system demonstrated the presence of a dark body with 36 years around one of the stars in the 1890s. A work by Forest Ray Moulton, However, demonstrated how unstable a three-body system would be given those orbital characteristics. Peter van de Kamp of Swarthmore College produced another well-known set of detection claims in the 1950s and 1960s, this time for planets around Barnard's Star. Most early claims of detection are now considered false by astronomers.

Using pulsar timing differences, Andrew Lyne, M. Bailes, and S. L. Shemar claimed to have found a pulsar planet orbiting PSR 1829-10 in 1991.

The assertion momentarily attracted a lot of attention, but Lyne and his group quickly disavow it.

Verified findings

The Extrasolar Planets Encyclopedia lists 5,108 verified exoplanets as of 1 July 2022, including a couple that was confirmations of contentious claims from the late 1980s.

The Canadian astronomer's Bruce Campbell, G. A. H. Walker, and Stephenson Yang of the Universities of Victoria and British Columbia made the first reported finding that was subsequently confirmed in 1988. Their radial-velocity studies revealed that a planet circles the star Gamma Cephei, even if they were hesitant to declare a planetary discovery. Astronomers were dubious about these and other related findings for several years, in part because the observations were made at the extreme edge of instrumentation at the time.

Some of the apparent planets could have been brown dwarfs, which are objects that fall between planets and stars in terms of mass. Additional findings that indicated the presence of the planet circling Gamma Cephei were reported in 1990, but following research in 1992 once again raised major concerns. Finally, in 2003, new technologies made it possible to establish the planet's existence.

Aleksander Wolszczan and Dale Frail, two radio astronomers, reported the finding of two planets circling the pulsar PSR 1257+12 on January 9, 1992.

MiIt was confirmed that this discovery was made, and it is widely accepted that this was the first confirmed finding of an exoplanet. These findings were confirmed by other observations, and the discovery of the third planet in 1994 reignited public interest in the subject.

These pulsar planets may have formed in a second round of planet formation from the peculiar byproducts of the supernova that gave rise to the pulsar, or they may have developed from the rocky cores of gas giants that somehow managed to survive the supernova and then decayed into their present orbits. At the time, it was thought implausible that a planet could develop in the orbit of a pulsar due to the aggressive nature of these stars.

Astronomers headed by Donald Backer concluded that a third item was necessary to account for the detected Doppler shifts while analyzing what they believed to be a binary pulsar (PSR B1620−26 b). The gravitational effects of the planet on the pulsar and white dwarf's orbits were measured within a few years, allowing for an estimate of the mass of the third object, which was too tiny to be a star. Stephen Thorsett and his colleagues revealed their discovery that the third object was a planet in 1993.

The neighboring G-type star 51 Pegasi was the exoplanet that Michel Mayor and Didier Queloz of the University of Geneva first definitively identified as circling a main-sequence star on October 6, 1995.

The Observatoire de Haute-discovery, Provence's which launched the current exoplanetary discovery period, was honored with a portion of the 2019 Nobel Prize in Physics. The quick discovery of several new exoplanets was made possible by technological advancements, particularly in high-resolution spectroscopy. Astronomers discovered exoplanets indirectly by observing their gravitational effects on the motion of their host stars. Later, by tracking changes in a star's apparent brilliance when an orbiting planet passed in front of it, more extrasolar planets were found.

The majority of exoplanets that were first discovered were large planets with tight orbits around their parent stars. These "hot Jupiters" startled astronomers since massive planets should only develop far from stars, according to planetary formation theories.

But when additional planets of different types were discovered throughout time, it became apparent that hot Jupiters were a minority among exoplanets. Upsilon Andromedae was discovered to contain several planets in 1999, making it the first main-sequence star to do so. The first planet to be found in an orbit around a pair of main-sequence stars is Kepler-16.

The Kepler Space Telescope, operated by NASA, found 715 newly validated exoplanets orbiting 305 stars on February 26, 2014. Verification by multiplicity, a statistical method, was used to examine these exoplanets. The majority of verified planets were formerly gas giants the size of Jupiter or bigger because of their easier detection, whereas the Kepler planets are mostly between Neptune and Earth in size.

On July 23, 2015, NASA revealed Kepler-452b, a nearly Earth-sized planet orbiting a G2-type star's habitable zone.

Constellation Virgo.

This exoplanet, Wolf 503b, was found orbiting an "Orange Dwarf" star type and is twice as big as Earth. Wolf 503b is so near to the star that one orbit may be completed in as little as six days. The only exoplanet of the size that can be located close to the so-called Fulton gap is Wolf 503b.

The finding that it is uncommon to locate planets within a given mass range is known as the Fulton gap, which was originally made in 2017. This expands the scope of research for astronomers who are now determining whether planets discovered in the Fulton gap are gaseous or rocky.

The first Earth-sized planet in the habitable zone spotted by TESS was TOI 700 d, which was discovered in January 2020, according to experts.

Methods of detection

Digital imaging

When compared to their parent stars, planets are quite dim. A Sun-like star, for instance, is around a billion times brighter than the light reflected from any exoplanet circling it. Such a dim light source is difficult to identify, and the parent star's glare makes it more difficult yet.

To eliminate glare and preserve the capacity to detect the light from the planet, it is required to block the light from the parent star. This presents a significant technological difficulty and calls for great thermal stability. All directly observable exoplanets are huge (massively more massive than Jupiter) and far from their parent star.

Many gas giants will be imaged by specially constructed direct-imaging telescopes like

Gemini Planet Imager, VLT-SPHERE, and SCExAO, although the great majority of known extrasolar planets have only been discovered indirectly. The indirect techniques that have worked well include the ones listed below:

indirect techniques

The visible brightness of a star dims somewhat when a planet transits (or passes) in front of the disk of its parent star. The size of the star and the size of the planet are two parameters that affect how much the star dims.

The likelihood that an exoplanet in a randomly oriented orbit will be detected to transit the star is rather low since the transit technique requires that the planet's orbit cross a line of sight between the host star and Earth. This approach was adopted by the Kepler telescope.

Doppler technique or radial velocity

The star goes in its own little orbit around the system's mass center like a planet circles a star. The Doppler effect causes shifts in the star's spectral lines, which may be used to measure changes in the radial velocity, or the rate at which the star is moving towards or away from Earth. One m/s or even somewhat lower fluctuations in radial velocity may be seen.

Variation in transit time (TTV)

Each planet somewhat messes with the orbits of the others when there are numerous planets. Thus, little fluctuations in a planet's transit timings might point to the existence of another planet, which may or may not be transiting. For instance, fluctuations in the transits of Kepler-19b's planet raise the possibility that Kepler-19c, which is not currently transiting, is a second planet in the system.

Transit time variability (TDV)

An animation demonstrating how the time of planet transits in one- and two-planet systems differs

A planet's transit duration might vary greatly from transit to transit if it circles more than one star or has moons. This technique hasn't led to the discovery of any new planets or moons, but it has been used to successfully confirm a lot of circumbinary planets that are transiting. [86]

The use of gravitational lenses

When a star's gravitational field behaves like a lens, amplifying the light of a far-off background star, the phenomenon is known as microlensing. The magnification may exhibit observable irregularities over time due to planets circling the lensing star.

The microlensing approach is particularly sensitive to identifying planets between 1 and 10 AU from Sun-like stars, in contrast to most other methods that have a detection bias towards planets with tiny (or for resolved imaging, big) orbits.

Astrometry

Astrometry is taking exact measurements of a star's location in the sky and tracking how that location changes over time. It may be possible to see a star's velocity as a result of a planet's gravitational pull. But this strategy has not yet proven particularly effective since the motion is so minute.

Only a handful of its detections have been called into question, but it has been effectively used to research the characteristics of planets discovered in other methods.

Pulse duration

A pulsar, which is a tiny, ultradense star that has had a supernova explosion, revolves while continuously emitting radio waves.

The timing of the pulsar's measured radio pulses will exhibit tiny irregularities if planets circle it.

Using this technique, the first extrasolar planet was officially discovered. However, as of 2011, only five planets orbiting three distinct pulsars have been discovered using this method.

Fluctuating star time (pulsation frequency)

There are various more kinds of stars that display periodic activity, similar to pulsars. A planet circling it may sometimes create variations in the periodicity. By using this technique, a few planets have been found as of 2013.

Modulations of reflection and emission

A planet absorbs a significant quantity of sunlight when it circles a star at a relatively near distance. Since planets seem to have phases from Earth's perspective or shine more brightly from one side than the other due to temperature variations, the quantity of light varies as the planet moves around the star.

Radiative beaming

Relativistic beaming counts the star's motion-induced flux that is being viewed. As the planet approaches or recedes from its host star, the star's brightness varies.

Changes in ellipsoidal form

Massive planets may somewhat alter the shape of their host stars if they are near enough. Because of this, the brightness of the star varies significantly depending on how it rotates in relation to Earth.

Polarimetry

A polarized light reflected from the planet is distinguished from an unpolarized light emitted from the star using the polarimetry technique. Although a few previously known planets have been recognized, no new planets have been found using this technique.

Disks circling stars

Many stars are surrounded by disks of cosmic dust, which are assumed to be the result of comet and asteroid impacts.

Because the dust absorbs starlight and reemits it as infrared radiation, it may be seen. Though not regarded as a certain detection tool, features in the disks may hint towards the existence of planets.

Evolution and formation

A few to tens of millions of years after the birth of their star, planets may begin to form.

The planets of the Solar System can only be seen in their present configuration, but we are able to see planets at various phases of development by observing other planetary systems with differing ages.

The range of observations includes planetary systems older than 10 Gyr and young proto-planetary disks where planets are actively forming. In a gaseous protoplanetary disk, planets develop by forming hydrogen/helium envelopes.

Depending on the mass of the planet, these envelopes cool and shrink with time, and ultimately part or all of the hydrogen and helium escapes into space. This implies that, if they develop early enough, even terrestrial planets might have huge initial radii.

 As an example, consider Kepler-51b, which has just approximately twice the mass of Earth but is virtually Saturn's size, despite having a mass 100 times that of Earth. With just a few hundred million years old, Kepler-51b is quite young.

planet-supporting stars

Main text: Stars that support planets

The spectral categorization used by Morgan-Keenan

Exoplanet depicted by an artist around two stars.

In general, each star has at least one planet.

 There is a "Earth-sized"[b] planet in the habitable zone for around 1 in 5 Sun-like stars.

The majority of known exoplanets revolve around main-sequence stars in the spectral ranges F, G, or K, which are approximately comparable to the Sun. Less planets with a mass large enough to be identified using the radial-velocity approach are expected to exist around lower-mass stars (red dwarfs, of spectral group M).

Despite this, the Kepler mission, which employs the transit technique to locate smaller planets, has found many tens of planets orbiting red dwarfs.

A link between a star's metallicity and the likelihood that it is home to a large planet the size of Jupiter has been discovered using Kepler data. Giant planets are highly likely to exist in stars with greater metallicities than in stars with lower metallicities.

A few circumbinary planets have been found that orbit both members of a binary star system, while other planets circle just one component of a binary star system. A few planets in triple star systems including one in the quadruple Kepler-64 system are known.

stars that support planets

In general, each star has at least one planet.

There is a "Earth-sized"[b] planet in the habitable zone for around 1 in 5 Sun-like stars.

The majority of known exoplanets revolve around main-sequence stars in the spectral ranges F, G, or K, which are approximately comparable to the Sun. Less planets with a mass large enough to be identified using the radial-velocity approach are expected to exist around lower-mass stars (red dwarfs, of spectral group M).

Despite this, the Kepler mission, which employs the transit technique to locate smaller planets, has found many tens of planets orbiting red dwarfs.

A link between a star's metallicity and the likelihood that it is home to a large planet the size of Jupiter has been discovered using Kepler data. Giant planets are highly likely to exist in stars with greater metallicities than in stars with lower metallicities.

A few circumbinary planets have been found that orbit both members of a binary star system, while other planets circle just one component of a binary star system. A few planets in triple star systems including one in the quadruple Kepler-64 system are known.

Chapter 3

5 Ways To Discover a Planet

Radiometric Velocity

936 planets were found while looking for wobble.
Astronomers may see a change in the hue of light due to the motion of stars caused by planets in their orbits.

SIZE RULES

Imagine a game of tug-of-war to help understand the gravitational interaction between a planet and a star. The star, a huge object with a very strong gravitational field, is on one side.

The planet is significantly smaller and has far less gravity on the opposite side.

The star is the clear winner in this match. Planets orbit stars rather than the other way around for this reason.

The planet possesses some gravitational pull, despite its modest size. Even while the impact on the host star is far less significant than the impact on the planet, it nevertheless affects it.

The gravity game can only be played by two.

Look at the animation in the image above. The situation seems to be normal at first. A giant star and a little planet are present, and the small planet revolves around the big star. This has undoubtedly been seen by you a lot.

But take a look at the star. Observe how it is also moving a little bit. Although this animation exaggerates the impact, it is what truly occurs in space. The star 'wobbles' around a small amount as a result of the planet's gravity.

As you may expect, a planet's impact on its star increases with size. Small planets like Earth barely slightly jiggle their stars. The impact of larger planets like Jupiter is substantially greater.

We may infer information about a star's planets from its "wobble," including the number, size, and the number of them.

ANALYSIS OF THE DOPPLER DATA

Finding exoplanets is made much easier by wobbling stars, but how can we spot them?

The technique is known as a "Doppler shift." It has the name of the scientist who discovered it almost 150 years ago.

Energy travels in waves, including sound, radio waves, heat, and light. similar to the waves in the animation up top.

Those waves may be extended or compressed depending on how the item creates its moves.

You may not be aware of it, yet you have undoubtedly already encountered the Doppler effect. Have you ever noticed how an ambulance approaching you on the street makes a higher-pitched sound before lowering it as it rushes away?

The reason is that the waves congregate and squeeze together when an energy-emitting object—such as a large, blazing star or an ambulance speaker—gets closer to you. The waves also enlarge as the thing moves away.

These wavelength shifts alter how we interpret the energy that we are seeing or hearing. The pitch of sound waves rises when they compress together. Additionally, visible light waves seem bluer in hue as they converge.

A lower pitch is produced when sound waves are stretched out. Additionally, when visible light waves elongate, an object seems more reddish.

Scientists may use redshift, a change in hue, to determine whether an object in the sky is moving closer to us or further away from us.

Bringing everything together
The animation shown above demonstrates how this technique works. As the planet swings back and forth, the star's orbit is disturbed, which causes the light waves to compress and then stretch, altering the hue of the light that humans can see.

HIGHLY SUCCESSFUL

One of the earliest strategies to discover exoplanets that were also one of the most fruitful is the radial velocity approach. This procedure is often used as an additional step to validate planets discovered using other techniques.

The Keck Telescopes in Hawaii and the La Silla Observatory in Chile are two major observatories where this study is conducted. Many scientists and telescopes all over the globe employ this approach to find exoplanets.

SEARCHING FOR SHADOWS IN TRANSIT

A planet may be measured as dimming the light of its star by passing directly between it and the observer.
One of the most fascinating astronomical phenomena you will ever see is a solar eclipse. It occurs when the moon squarely faces the sun, obstructing the sun's light.

Similar to how the transit technique identifies exoplanets, this. A planet blocks part of the light from the star it orbits when it is directly between the observer and the star. That star does, for a moment, become fainter. Even if it's a little alteration, it's enough to let scientists know that an exoplanet is orbiting a far-off star.

Astronomers refer to the graph that is being generated on the left side of the video as a "light curve." It is a graph showing the amount of light coming from the star. The light curve shows this decrease in brightness that occurs when a planet moves in front of a star and partially blocks some of its light.

SIZE SENSE

We can learn a lot about the planet generating a transit from its size and duration. Deeper light curves are produced by larger planets because they block more light. When you click "various planet sizes and distances," which is highlighted in the animation above, you may see it.

A planet takes longer to circle and pass in front of its star the farthest it is from its star. So, the farther a planet is from its star, the longer a transit event lasts.

MULTIPLE SIGNALS

You can observe that light curves get more convoluted when many planets are transiting a star by selecting "multiple planets" from the drop-down menu. Even while it requires more effort on the part of the astronomers to identify each planet in the data, the combined light curves may provide us with the same information as a single one.

ATMOSPHERIC INTENTIONS
The transit technique is helpful for more than only locating planets; it may also provide data on a planet's temperature or atmosphere's makeup.

Some of the starlights enters an exoplanet's atmosphere as it passes in front of its star.

Scientists can deduce important details about the makeup of this light by examining the colors it emits. They've discovered everything on other planets using this technique, from methane to water vapor.

KNOWLEDGE GOLDMINE

Finding new exoplanets has proven very effective using the transit approach. From 2009 to 2013, NASA's Kepler spacecraft used the transit technique to search for planets. It detected hundreds of potential exoplanet discoveries and provided scientists with useful data on the distribution of exoplanets in the galaxy.

DIRECT IMAGING Image-making

By reducing the glare of the stars they circle, astronomers may snap photos of exoplanets.

Exoplanets are millions of times fainter and further distant than the stars they orbit. Therefore, it should come as no surprise that photographing them in the same manner as photographing, say, Jupiter or Venus, is quite difficult.

It is being accomplished via new methods and quickly increasing technology.

Because the stars they orbit are millions of times brighter than their planets, astronomers have a difficult time directly seeing exoplanets. The enormous volumes of radiation emanating from the planet's host star overwhelm any light reflected off the planet or heat radiation from the planet itself. It's comparable to attempting to discover a firefly darting around a spotlight or a flea in a lamp.

SUNLIGHT BLOCKERS

On a sunny day, you could use your hand, a pair of sunglasses, the sun visor of your automobile, or another object to obscure the sun's brightness so you can see other things.

The devices made to directly observe exoplanets work on the same principles. To obstruct the brightness of stars that potentially have planets around them, they use a variety of strategies. They will be able to see potential exoplanets more clearly after the star's glare is diminished.

HOW TO BUILD A LIGHT-BLOCKER

A star's light may be blocked using one of two basic techniques, according to scientists.

One method, known as coronography, employs a component within a telescope to obstruct a star's light before it reaches the instrument's detector. Coronagraphs, which are constructed as internal extensions for telescopes, are currently employed by ground-based observatories to directly observe exoplanets.

Another approach is to use a "starshade," a piece of equipment that is placed to obstruct a star's light before it ever reaches a telescope. A starshade would be a separate spacecraft created specifically to place itself at the ideal distance and angle to obscure illumination from the star that scientists were watching for an exoplanet-hunting telescope.

THE FUTURE'S WAY

Although direct imaging is still in its infancy as an exoplanet-finding technique, there is considerable anticipation that it will someday become a crucial tool for discovering and identifying exoplanets. Future direct-imaging telescopes could be able to capture images of exoplanets that show their seas, landmasses, and atmospheric patterns.

MOVEMENT MICROLENSING

A Gravity Lens's Light

As a planet passes in front of a far-off star and Earth, gravity bends and concentrates light from the star.

Albert Einstein rethought the idea of gravity, seeing it less as a mystical affinity between things and more as a geometric characteristic of spacetime, among his many other breakthroughs.

Large things, in other words, stretch the boundaries of space. When a big object, such as a star or planet, has its gravity impacted by this action, light is forced to change its shape and direction.

CONVERGING TO LIGHT

Some fairly fascinating things may result from this course shift. Gravity may sometimes concentrate and bend light like a lens in a pair of glasses or a magnifying glass.

When a star or planet's gravity concentrates the light of another, more distant star in a manner that briefly makes it seem brighter, this is known as gravitational microlensing.

You can see how the light from the more distant star bends around the exoplanet in the video above, then around the star of the exoplanet. The gravity of the planet and the star concentrate the light rays of the distant star onto the observer like how a magnifying glass may focus the sun's light into a small, very brilliant spot on a sheet of paper.

The graph on the left shows how the brightness of the faraway star varies when its light is focussed upon the viewer using lenses. The star begins to get brighter, followed by a fleeting burst of brilliance caused by the planet's lensing activity.

After the planet is lensed, the light levels decrease but then rise again due to the star's ongoing lensing activity. The more distant star's brightness decreases as soon as it shifts out of the lensing star's ideal location.

A FLASH OF LIGHT THAT FLEW AWAY

An astronomer would compare a lensing event to a faraway star that gradually becomes brighter over a month or two before fading away. In the course of this brightening and fading process, a planet that is lensed appears as a fleeting blip of light.

Astronomers are unable to foretell the timing or location of these lensing occurrences. As a result, they must spend a lot of time scanning a vast portion of the sky. They examine the data to determine the approximate size of the star when they see a star becoming brighter and then fading in the pattern of lensing objects.

Astronomers sometimes record brief microlensing events that are caused by planets that are free-floating in space and do not revolve around a star. These occurrences offer us an indication of the prevalence of these 'rogue' planets in the cosmos. What that type of incident seems like through a telescope is seen in the animation to the right.

Minuscule Movements in Astronomy

Concerning other neighboring stars in the sky, a star may wobble in space due to the orbit of a planet.

For a description of how planets influence their stars to sway in space, first read the Radial Velocity text. Let's wait!

Astronomers can identify stars that are swaying because of the gravity of their planets in addition to using doppler shifts. The star's apparent location in the sky may shift, making the wobble obvious.

In other words, the movement of the star's location in space may be felt by scientists.

This technique, referred to as astrology, is still very challenging. It is highly difficult to precisely measure the wobbling from planets, particularly tiny ones the size of Earth since stars wobble at such a minute distance.

Scientists take a series of pictures of a star and some of the nearby stars in the sky to follow the motion of these stars. They compare the separations between these reference stars and the star they are looking for exoplanets in each image.

Astronomers may examine a star's motion concerning other stars to determine whether an exoplanet has been detected.

Because of how our atmosphere bends and distorts light, astrometry needs incredibly accurate optics and is particularly challenging to perform from the surface of the Earth.